STARK LIBRARY JAN - - 2022

DISCARD

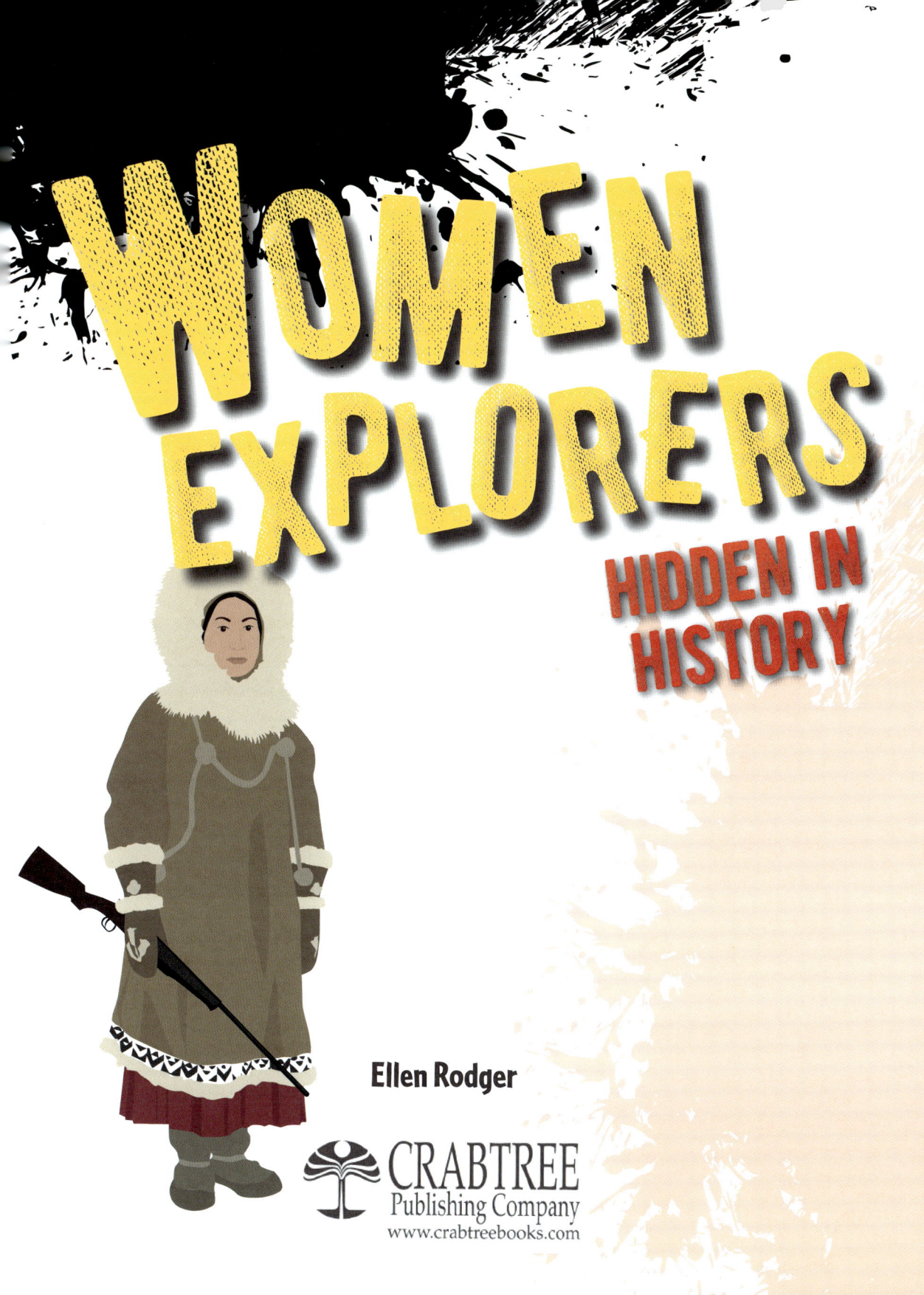

WOMEN EXPLORERS

HIDDEN IN HISTORY

Ellen Rodger

CRABTREE
Publishing Company
www.crabtreebooks.com

Author: Ellen Rodger

Editorial director: Kathy Middleton

Editors: Sarah Eason, Melissa Boyce

Proofreaders: Jennifer Sanderson, Elizabeth DiEmanuele

Design: Paul Myerscough and Jessica Moon

Cover design: Emma DeBanks and Jessica Moon

Illustrations: Jessica Moon

Photo research: Rachel Blount

Production coordinator and Prepress technician: Margaret Amy Salter

Print coordinator: Katherine Berti

Written, developed, and produced by Calcium

Photo Credits:
t=Top, c=Center, b=Bottom, l= Left, r=Right

Inside: Harry Palmer: p. 34b; Shutterstock: AAR Studio: p. 42b; Ersin Alok: p. 19t; Valery Bareta: p. 7; Bjul: p. 29b; Rafal Cichawa: p. 13; Rob Crandall: p. 24; Ace Diamond: p. 30b; Gustavo Frazao: p. 33; Aleksander Hunta: p. 23; Incredible Arctic: p. 36b; Ovchinnikova Irina: p. 27t; Zoran Karapancev: p. 22b; KH-Pictures: p. 20t; Lamnee: p. 30t; Kevin M. McCarthy: p. 39; Moroz Nataliya: p. 22t; Neveshkin Nikolay: p. 6b; Joy Prescott: p. 37; Rook76: p. 45b; Vaclav Sebek: p. 41; Snehit: p. 31; THPStock: p. 21; Pawika Tongtavee: p. 25b; Wikimedia Commons: p. 34t; after G. Browne/Wellcome Images: p. 45t; R. Carrick after Lieutenant James Rattray/Wellcome Images: p. 10; Daderot: p. 17; Megan Em: p. 5; Felipe Augusto Fidanza: p. 32l; William Hodges: p. 9t; Bilibin Ivan: p. 18; C. Jongen: p. 6t; Jan Kneschke: p. 9b; Jaan Künnap: p. 27b; Henry Maull: p. 40l; Unknown upload by Adrian Michael: p. 43; Edgar Samuel Paxson: p. 29t; RIA Novosti archive, image 67418/Alexander Mokletsov/CC-BY-SA 3.0: p. 15; Wellcome Images: p. 12; Shutterstock: p. 16b

Library and Archives Canada Cataloguing in Publication

Title: Women explorers : hidden in history / Ellen Rodger.
Names: Rodger, Ellen, author.
Description: Series statement: Hidden history | Includes index.
Identifiers: Canadiana (print) 2020015320X |
  Canadiana (ebook) 20200153218 |
  ISBN 9780778772965 (hardcover) |
  ISBN 9780778773047 (softcover) |
  ISBN 9781427124753 (HTML)
Subjects: LCSH: Women explorers—Biography—Juvenile literature. | LCSH: Women adventurers—Biography—Juvenile literature. | LCSH: Women explorers—History—Juvenile literature. | LCSH: Women adventurers—History—Juvenile literature.
Classification: LCC G200 .R63 2020 |
  DDC j910.92/52—dc23 | 910.92/52—dc23

Library of Congress Cataloging-in-Publication Data

Names: Rodger, Ellen, author.
Title: Women explorers : hidden in history / Ellen Rodger.
Description: New York : Crabtree Publishing Company, [2020] |
  Series: Hidden history | Includes index.
Identifiers: LCCN 2019054365 (print) | LCCN 2019054366 (ebook)
  ISBN 9780778772965 (hardcover) |
  ISBN 9780778773047 (paperback) |
  ISBN 9781427124753 (ebook)
Subjects: LCSH: Women explorers--Biography--Juvenile literature. | Women adventurers--Biography--Juvenile literature.
Classification: LCC G200 .R63 2020 (print) | LCC G200 (ebook) |
  DDC 910.92/52--dc23
LC record available at https://lccn.loc.gov/2019054365
LC ebook record available at https://lccn.loc.gov/2019054366

Crabtree Publishing Company
www.crabtreebooks.com   1-800-387-7650

Printed in the U.S.A./022020/CG20200102

Copyright © **2020 CRABTREE PUBLISHING COMPANY**. All rights reserved. No part of this publication may be reproduced, stored in a retrieval system or be transmitted in any form or by any means, electronic, mechanical, photocopying, recording, or otherwise, without the prior written permission of Crabtree Publishing Company. In Canada: We acknowledge the financial support of the Government of Canada through the Canada Book Fund for our publishing activities.

Published in Canada
Crabtree Publishing
616 Welland Ave.
St. Catharines, Ontario
L2M 5V6

Published in the United States
Crabtree Publishing
PMB 59051
350 Fifth Avenue, 59th Floor
New York, New York 10118

Published in the United Kingdom
Crabtree Publishing
Maritime House
Basin Road North, Hove
BN41 1WR

Published in Australia
Crabtree Publishing
Unit 3 - 5 Currumbin Court
Capalaba
QLD 4157

Contents

Chapter 1
History's Hidden Heroines........................... 4

Chapter 2
Epic in Europe.. 8

Chapter 3
Awesome in Asia..................................... 18

Chapter 4
Amazing in the Americas 28

Chapter 5
Acing It in Africa..................................... 40

Glossary.. 46
Learning More 47
Index and About the Author 48

CHAPTER 1
HISTORY'S HIDDEN HEROINES

There is no doubt that women have always been a part of journeys of discovery. Women have been adventuring since the earliest times, from Viking women crossing the ocean to find new lands to women traders traveling in caravans to transport goods through Africa. But there is little mention of them in most history books. Sadly, women have been largely ignored and written out of stories of early exploration.

Looking up the Heroines

Historians are only now beginning to uncover some of the fascinating tales of women adventurers, such as the women who sailed with the Dutch East India Company in the 1700s—while disguised as men! More recently, British adventurer Rosie Swale Pope proved that a woman going it alone can complete the longest around-the-world run in history and live to tell and write about it. She ran a total of 20,000 miles (32,187 km)!

Trading caravans with up to 1,000 camels crossed the Sahara between the Mediterranean and West Africa from 700 C.E. to 1500 C.E. There's no doubt women were a part of these caravans.

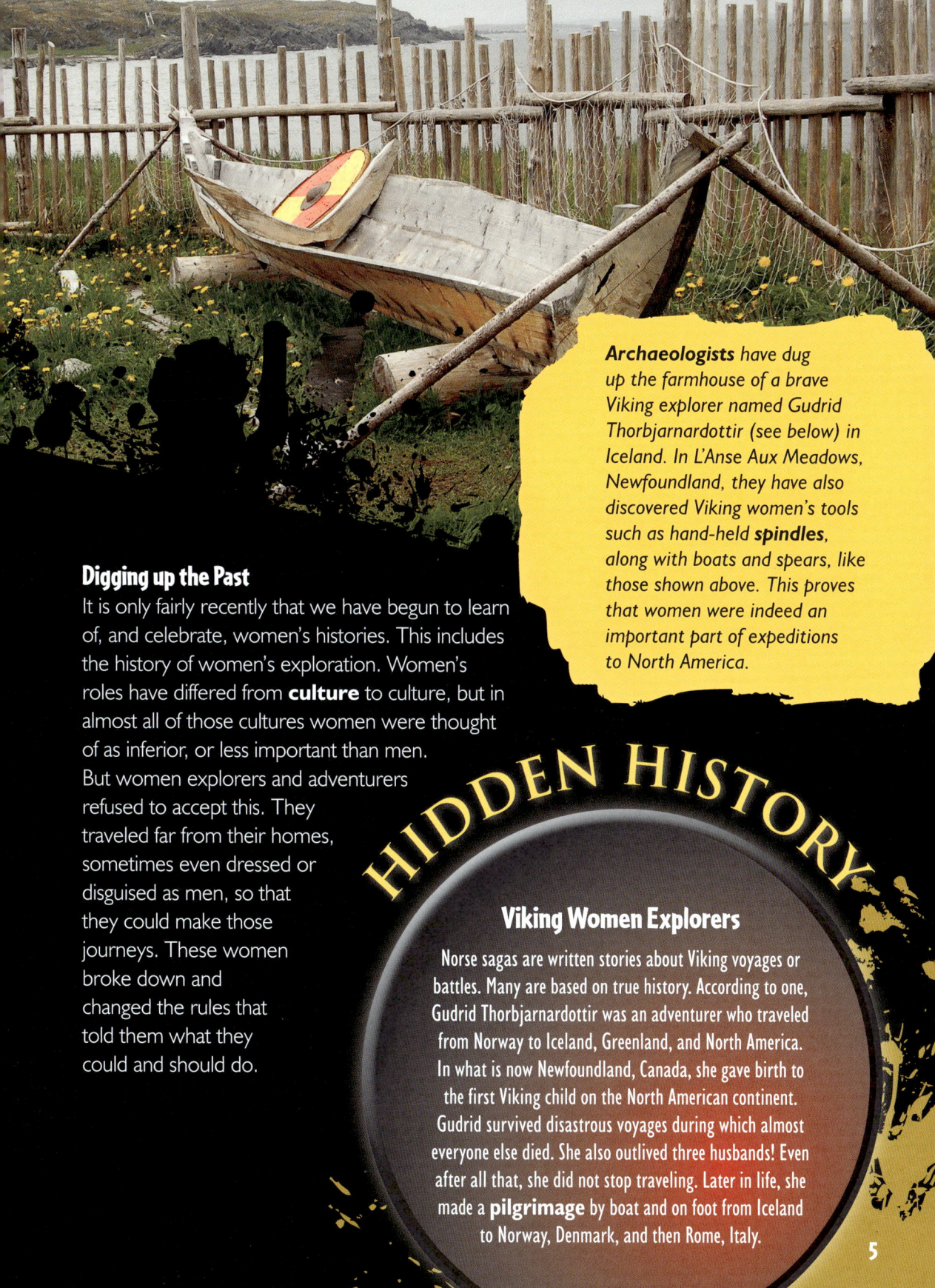

Archaeologists have dug up the farmhouse of a brave Viking explorer named Gudrid Thorbjarnardottir (see below) in Iceland. In L'Anse Aux Meadows, Newfoundland, they have also discovered Viking women's tools such as hand-held **spindles**, along with boats and spears, like those shown above. This proves that women were indeed an important part of expeditions to North America.

Digging up the Past

It is only fairly recently that we have begun to learn of, and celebrate, women's histories. This includes the history of women's exploration. Women's roles have differed from **culture** to culture, but in almost all of those cultures women were thought of as inferior, or less important than men. But women explorers and adventurers refused to accept this. They traveled far from their homes, sometimes even dressed or disguised as men, so that they could make those journeys. These women broke down and changed the rules that told them what they could and should do.

HIDDEN HISTORY

Viking Women Explorers

Norse sagas are written stories about Viking voyages or battles. Many are based on true history. According to one, Gudrid Thorbjarnardottir was an adventurer who traveled from Norway to Iceland, Greenland, and North America. In what is now Newfoundland, Canada, she gave birth to the first Viking child on the North American continent. Gudrid survived disastrous voyages during which almost everyone else died. She also outlived three husbands! Even after all that, she did not stop traveling. Later in life, she made a **pilgrimage** by boat and on foot from Iceland to Norway, Denmark, and then Rome, Italy.

5

So, Why Is Women's History Hidden?

Women's history is not as straightforward as men's history. The stories of women's lives and achievements have had to be pieced together from information hidden in forgotten ancient books or overlooked **records** and letters.

Women's roles, from ancient times until recent times, were very controlled. Women were supposed to be wives and mothers, and few had other options. For so long, women were not allowed a good education. Most ordinary women could not read or write. When you cannot write your own stories, your history is then told by others. Because men wrote much of history, the achievements of women were often ignored or removed.

Women, Why Write About Them?

The earliest mention of women explorers comes from the works of the Roman poet Virgil who was born in 70 B.C.E. Even then, the account was of a queen. In Virgil's account, the queen's name was Dido and she traveled from her home in Tyre, in Lebanon, to the island of Cypress, and then to North Africa. There, Queen Dido founded, or set up, the ancient trading city of Carthage.

This is a sculpture of Dido. Some writings say the name Dido means wanderer. The legendary queen traveled across the Mediterranean to set up the city of Carthage.

Some women explorers wrote about their travels in journals that they kept during their adventures. These records teach us more about these amazing women.

Carthage was once a great and wealthy city that traded with people all over the Mediterranean and beyond. Today, the ruins of Carthage can be seen on a large body of water called the Lake of Tunis in what is now Tunis, Tunisia, in North Africa.

Some historians think that Virgil's story is true and based on a real queen. Others believe it is make-believe. We don't know for sure, because Dido didn't tell her own tale. It was not until the 1700s and 1800s that women wrote their own stories of exploration in letters, journals, and books. A lot of what we know about women explorers comes from these works. These are their own words, describing their own adventures.

Exploring Without Respect

The history of all exploration is filled with **prejudice** about **race**, culture, and **gender**. Although they were groundbreakers, some women explorers also held the same views about race and culture as the men that wrote history. These views considered the white European culture to be better than Asian, African, or **Indigenous** North American cultures. The adventurers who held these beliefs did not respect the people who lived in the areas they explored—even though they needed the support and guidance of these people in order to survive their journeys.

CHAPTER 2
EPIC IN EUROPE

So, how did women explore when they were barely allowed to go out without a man? Well, they "wore the pants"–quite literally! At a time when women (outside of south Asia) wore only skirts, many of them had to disguise themselves as men.

Jean Baré, Mistress of Disguise

From housekeeper to scientist and world traveler, Jean Baré (or Baret) was a rule-breaker. She became the first woman to sail around the world, on a French expedition, from 1766 to 1769. She did all of this while disguised as a man! Born in Burgundy, France, Baré was an **orphan** by her teens, but found work as a housekeeper and nurse to the **naturalist** Philibert Commerson.

Coming Clean

Philibert was sickly, so Jean began to help him with any physical tasks his work required. She quickly became an important work companion. As the pair grew closer, Philibert decided to take Jean with him on an overseas trip. In 1766, they set sail for Uruguay, Brazil, Tahiti, and Mauritius. Jean disguised herself as a man to travel, because women were not allowed on voyages of exploration. However, when the ship arrived in Tahiti, the Tahitians quickly pointed out to the

Jean traveled as Philibert's valet, or personal manservant. She dressed in male clothing to look the part.

crewmembers that Jean was a woman! She and Philibert were allowed to sail to the Indian Ocean island of Mauritius, but no farther. Philibert died there soon after and Jean sailed home to France in 1774, completing her around-the-world trek.

Jean's expedition to Mauritius included three ships and was the first around-the-world expedition to include scientists.

Philibert named two tropical flowers after Jean. The Commerson's dolphin (shown below) is named after him.

HIDDEN HISTORY

Secret Child

Historians think that Jean and Philibert had a child together before they set off on their explorations. A "certificate of pregnancy" dated 1764 was filed by Jean in a town close to where she lived. The certificate was required by the French government for women who became pregnant outside of marriage. The child was placed in an **orphanage** and died a year later. Although Jean and Philibert never married, Philibert did leave her money and furniture from their Paris apartment after he died.

9

Lady Hester Stanhope, Queen of the Desert

A shipwreck, younger boyfriends, and disapproving relatives—**Lady** Hester Stanhope faced them all. This tall, confident, and outspoken daughter of a British **earl** wasn't afraid to stand up to anyone who pushed her around. That served her well when she set off on traveling adventures throughout Europe and the Middle East.

Busting Loose

Lady Hester was born in 1776, a time when most women, even the daughters of British noblemen, had little education. That didn't keep Hester from longing for a different life. As a teenager, she angered her father because of her wild ways. They did not get along, so Hester left home to live with her uncle—the British prime minister. She organized his household and when he died, she was given a large sum of money for her work. Hester used the money to break loose and travel the world!

Lady Hester was fascinated by the land and people of the Middle East, and especially those who lived in the desert.

HIDDEN HISTORY

A Hidden Woman

Shipwrecked on the way to Cairo, Egypt, Lady Hester lost all of her luggage. She simply replaced her heavy dresses with men's clothes, so she looked like a Turkish man. From then on, she wore men's clothing. She refused to change back into women's clothing and even shaved off her hair so that her **turban** fit better!

Boats and Camels

What at first was supposed to be a short, fun trip to the Mediterranean became an adventure of many years. Lady Hester traveled in style, sailing, riding horses, and even crossing deserts on a camel. But she didn't travel alone. She brought a doctor, a maid, a boyfriend, and a lot of luggage with her. The group traveled to Gibraltar in Spain, Athens in Greece, and Constantinople (now Istanbul) in Turkey. Then Lady Hester turned her attention to the Middle East.

The sight of a tall, pale, foreign woman wearing a turban, a cloak, and carrying a pistol and a sword shocked people. Why do you think Hester chose to dress as a man on her journeys?

Here Comes a Queen

Lady Hester became one of the first Europeans to travel the deserts of Syria and Lebanon. In Damascus, Syria, she was met by locals who thought she must be a queen to have traveled so far because she dressed and acted as she liked.

Later, Hester demanded local **tribesmen** travel with her camel caravan to the ancient city of Palmyra. As she entered the city, she was greeted by dancing women. A young girl was said to have placed a **palm wreath** on Hester's head. This really made Lady Hester think she was special—she started to believe she actually was royal! In her letters she even wrote, "I have been crowned Queen of the Desert."

A Queen Needs a Castle

Hester loved the attention she got on her travels and decided to make the desert her home. She rented a small **convent** in Lebanon and began acting like the local queen. When her boyfriend returned to Britain, she simply found another young man to take his place.

Lady Hester loved the attention she got when riding and acting like the queen of the desert. As a foreigner, Hester had a lot more freedom than local women.

Lady Hester was both generous and cruel. She took in strangers and beggars, but also punished the locals. After a European traveler was killed near her desert home, Hester ordered several villages to be burned. That she actually had the power to do so shows how forceful she was. Unfortunately, Lady Hester also then upset the local people by riding through their destroyed villages like a great, powerful warrior.

Today, the once-great city of Palmyra lies in ruins. People no longer live there.

ALONE IN THE DESERT

When her fellow travelers died or returned to Britain, Hester moved to an **abandoned monastery**. Lady Hester lived there with 30 cats. But her **debts** piled up and her British **pension** was canceled to pay for her bills. To complain, Hester wrote a letter to an actual queen—Britain's Queen Victoria. But she received no response. Hester then walled herself up in her monastery home and died there, alone.

Valentina Tereshkova, Reaching for the Stars

Valentina Tereshkova holds a title no other person can **claim**: the first woman in space. She also remains the youngest woman in space—when her *Vostok 6* rocket launched in 1963, she was just 26.

Racing to Space

In the 1950s and 1960s, the **Soviet Union** was caught up in a "space race" with the United States. The Soviets launched the first **satellite** in 1957. In 1961, they sent the first man into space—Yuri Gagarin. The United States hit back with a program to send **astronauts** to the Moon. During this back-and-forth competition, Valentina quietly trained to become a cosmonaut, which is an astronaut in the Soviet space program. At the time, it was the first program to train women for space flight.

*Valentina's history-making **orbit** earned her the title of Hero of the Soviet Union, and later the United Nations Gold Medal of Peace.*

14

Valentina married fellow cosmonaut Valery Bykovsky, who flew on three space missions. One of them was during Valentina's Vostok 6 flight. They came within 3 miles (5 km) of each other in space!

Loaded and Launched

As her *Vostok 6* spacecraft was launched on June 16, 1963, Valentina said "Hey sky, take off your hat. I'm on my way!" She spent 70.8 hours, or almost 3 days, in space, and orbited Earth 48 times. The *Vostok*'s tiny cabin was attached to a **service module** and Valentina did a number of tests that would help future Soviet cosmonauts. She landed back on Earth in a mountainous area near the **Kazakhstan border**. After climbing out of her capsule, she joined a group of locals for supper!

ADRENALINE RUSH

As a teenager, Valentina loved skydiving. She had a taste for adventure and became skilled as a parachutist, making more than 126 jumps. This hobby didn't go over well with her mother, but it helped get her into the Soviet space program— cosmonauts had to parachute from their **capsules** before they hit the ground. Valentina was one of a handful of women picked for the space program. During training, she showed that she could handle **zero gravity** and the loneliness of space. Of all the women in the program, she was the only one chosen for space flight.

Annie Londonderry, Biking Like a Boss

A fearless mother-of-three, Annie Londonderry knew the value of a good story. And she made her way across the world by peddling it—Annie biked around the world on a bet! Except there was no bet...

Selling a Story

Annie was a smart woman from Boston, Massachusetts, who had a thirst for adventure and fame. She decided to cycle around the world to achieve both. To raise funds for her trip, she told people that someone had made a bet with her, promising to give her money if she biked all the way around the world. But Annie had made up the bet story to get people interested in her trip! She knew that if her adventure became famous, companies would pay her to advertise their products, and that would pay for the trip.

Landing a Deal

Annie's real name was Annie Cohen Kopchovsky. In 1894, just before her trip, she changed her last name to Londonderry as part of a **sponsorship deal** with a bottled

Annie started her journey wearing skirts and **bloomers**, but soon switched to wearing more comfortable men's suits. She was ahead of her time. Many people complained that they thought dressing like a man was unladylike.

This is the route that Annie took to cycle around the world.

16

water company, the Londonderry Lithia Spring Company. The company gave her $100 to carry their name on her bike around the world. This may have been one of the first female sponsorships in sporting history.

Spunky Traveler

On June 25, 1894, Annie set off on her trip. Although she had little experience bicycling, she still rode from Boston to Chicago, then on to New York City, where she boarded a ship with her bike. Her first stop was France. From there, Annie traveled by steamship to Egypt, Sri Lanka, Singapore, Hong Kong, China, and Japan.

At each stop, Annie cycled through cities or the countryside, along the way stopping to give talks in which she told stories about meeting kings and queens or fighting tigers! From Japan, she sailed back to the United States, landing in San Francisco. She then traveled by both bike and train across the country, back to Chicago—all in under 15 months. In Chicago, she picked up her $10,000 prize. Then she went back to her family and began her next great adventure, as a journalist.

Bicycle riding was challenging work in the late 1800s. Bikes were not designed for comfort or for speed, making Annie's efforts even more impressive.

HIDDEN HISTORY

Were All of Annie's Stories True?

Like many explorers and adventurers, Annie sometimes made up stories about what happened on her travels, including hunting Bengal tigers in India (she didn't)! We can't be sure how many of the stories she told were really true, and how many were make-believe. The truth lies hidden in the past!

CHAPTER 3

AWESOME IN ASIA

Women explorers of the past were made of tough stuff! Exploring foreign territories without modern transportation, such as jet planes and modern maps, made traveling hard for anyone. But male explorers often had funding for their trips. Women, on the other hand, did not get the same support, which meant they often traveled by the seat of their skirts (or baggy pants).

Freya Stark, Rule Breaker

Inspired by reading *The Thousand and One Nights* as a sickly child, Freya Stark dreamed of adventure in the Middle East. She didn't set off on that adventure until she was in her late 30s, but she made up for her late start by traveling into her 70s. She often traveled alone.

The Thousand and One Nights tells of many wonderful and exciting adventures in Arabian lands!

The Gift of Gab

Born in France in 1893, Freya had an unusual childhood. She grew up in Italy and England mostly with her free-thinking mother and grandmother. She spoke French, Italian, and German, and never entered a school classroom until she went to college to study even more languages, including Arabic and Persian.

Freya was a nurse during World War I (1914 to 1918), and sometime after, set out on the first of many explorations of Persia (Iran), Turkey, Yemen, Afghanistan, India, Egypt, and Kurdistan.

Valley of the Assassins

Freya wanted to find the Valley of the **Assassins** that fellow traveler Marco Polo had described in his writings in the **medieval** period. The assassins were a group of warriors famous for their fighting skills and their magnificent castle in the desert. Freya didn't have a map showing where the castle was. But even that didn't stop her—she just found a local guide and set off!

Freya traveled on foot, camel, donkey, and sometimes by car on her journey. She spoke to local people on the way, who helped point her in the right direction of the castle. Eventually, Freya found the castle ruins near the Elburz Mountains in northern Persia. Freya wrote about her amazing journey in her first book, *Valley of the Assassins*.

Traveling through the Elburz Mountains was tough. Freya was a determined explorer and did it with one guide and little money.

HIDDEN HISTORY

Looking for Lost Cities

Freya often set out to find lost cities that had been hidden in history and were in places few people visited. Freya was known for her respect of the people she met during her travels. She tried to listen and learn from them. This was unusual for explorers of the time.

Freya Stark's last expedition was in 1968 in Afghanistan when she was 75. There she studied the **Minaret of Jam**, then published a book on it. In all, Freya wrote 24 books about her journeys. She died when she was 100 years old.

Alexandra David-Néel, Dirtbag Daredevil

In the mountain and rock-climbing world, "dirtbag" describes someone who loves climbing so much they give their whole life to it. These determined climbers often sleep in the wild and get really dirty so they can live out their rock-climbing dreams. Alexandra David-Néel has been described as the "original dirtbag" because she traveled mostly alone to far-flung places, climbed in the Himalayas, and even lived for a time in a cave!

Walking through the **Alps** to Lake Maggiore was no mean feat for a teen on her own, but Alexandra did it when young women were rarely allowed to travel without family.

Catching the Travel Bug

Paris-born Alexandra loved exploring new places and religions. She caught the travel bug early and made her first journey alone as a teenager in 1883. At just 16, she took off from her family's home in Belgium without telling anyone. Alexandra took a train to Switzerland where she walked across the Alps to Lake Maggiore, in Italy. The next year, Alexandra rode a bicycle through Spain with her bags tied to the handlebars!

Alexandra spent 14 years traveling and learning in India, Tibet, Japan, Korea, Mongolia, and China. In some of these places, she climbed where no other woman had ever climbed before.

NO ORDINARY MARRIAGE

In 1904, at the age of 36, Alexandra got married. But it wasn't an ordinary marriage. She spent most of it living apart from her husband. He knew she had the travel bug and in 1911, they agreed that she would spend 19 months exploring Asia. It would be another 14 years before they saw each other again! The couple kept in regular touch through letters. They remained married and Alexandra was deeply saddened when she received news of her husband's death in 1941.

Tell Me Everything!

Alexandra later became fascinated by **Tibetan Buddhism**, and it would become her passion. She did everything to learn more about it, including learning several languages, traveling to South Asia, and studying with guides.

Singing for Her Supper

Alexandra was all about adventure. As a young woman, she studied and worked as an opera singer. Her singing career saw her touring many places, including China, the Middle East, and North Africa. But her opera travels were not enough. Alexandra wanted to know the cultures of the places she traveled to. In India, she became a student of a Buddhist monk, learned to speak Tibetan, and to read the ancient **Hindu** language of Sanskrit.

Alexandra even visited the Potala Palace in Lhasa, which was the winter home of the **Dalai Lamas** from 1649 to 1959. She had to climb 12,100 feet (3,700 m) to reach it!

> Alexandra learned tummo, a breathing exercise that helped keep her warm. It helped her survive crossing the Himalayas over a 19,000-foot (5,791 m) pass in the middle of winter.

Free Spirit

Described as a "free spirit," Alexandra boldly pushed passed any obstacle in her way. She visited many Buddhist monasteries in her travels and made friends with several Buddhist monks, including the Dalai Lama. In the Indian state of Sikkim, in the eastern Himalayas, she became an advisor to the Maharaj, or ruler.

Through the Mountains

Surrounded by some of the highest mountains in the world, Tibet was a land of mysteries for centuries. Foreigners were discouraged and sometimes not allowed to travel there, particularly to the capital Lhasa. It was known as the forbidden city. This fact did not put Alexandra off. Disguised as a farmer and traveling with a young Tibetan guide, she trudged up mountain foothills in handmade boots, becoming the first European woman to travel to Lhasa.

> The Dalai Lama (right) is the spiritual leader of Tibetan Buddhism. Buddhism is a religion that teaches people to learn about themselves and their path through life. There have been 14 Dalai Lamas in the history of the religion.

Into Forbidden Lands

In 1916, when Alexandra crossed the border into Tibet, World War I was raging in Europe. Knowing that the Tibetan **authorities** were against her traveling into forbidden **territory** and that she could not return to Europe, Alexandra decided to travel farther into Asia. With her young Tibetan guide, she journeyed more than 5,000 miles (8,047 km) through Japan, Korea, Mongolia, and China. Alexandra traveled on foot, yak, donkey, and horse.

During her travels, Alexandra lived at a monastery for three years and **translated** ancient manuscripts. In 1924, disguised as a local, she slipped into the forbidden city of Lhasa. Alexandra wrote about her forbidden treks and her time learning about Tibetan Buddhism in several books. She was awarded a gold medal from the Geographical Society of France for her work. Alexandra **adopted** her Tibetan guide, who followed her back to France. She lived there until she was 101, but never lost her love of travel. In fact, she applied for a new passport just months before her death!

WRITING ABOUT HER WANDERS

Alexandra wrote about her adventures and the teachings of Buddhism in 30 books. Some of her last journeys were to China and eastern Tibet.

Alexandra traveled by donkey and yak (shown below) through China. She made it through Japan and Korea just before World War II (1939 to 1945) started.

Isabella Bird, Older and Bolder

Isabella Bird was born in 1831 in England. She traveled a lot in her early life, but she did even more exploring after she was 60 years old.

Ah, Travel, Good for the Health!

As a child, Isabella was often sick. She also had serious lifelong back pain that might have cut her traveling days short, if she hadn't been so stubborn. In fact, her family encouraged her travels, believing it was good for her health. Isabella set off on her first voyage of exploration at 23 years old.

Later Wanderings

Right from the start of her travels, Isabella wrote about what she saw and experienced, and published her stories in her books. She traveled all over the world, from Europe, Africa, and Asia to Australia and North America. In Hawaii, she climbed the volcanoes Mauna Kea and Launa Loa. She also rode on horseback for 3,000 miles (4,828 km) through the Rocky Mountains, and then stayed for months in a two-room cabin.

Isabella wrote about how ordinary people lived in the countries she visited. She was fascinated by everyday life, such as how people like these Moroccan women cooked.

KEEPING FREE

At a time when most women married fairly young and were expected to have children, Isabella did neither. Travel and adventures took up most of her adult life. She waited until she was 50 to marry (and even then, she wasn't too eager!), and she married a younger man. Isabella didn't want to give up her freedom and insisted that she would continue to travel on her own during her marriage. But she never did, even though her husband, Dr. John Bishop, supported her adventuring. When he died just five years after their wedding, she went back to the life she knew best—exploring.

She searched for the source, or start, of the Karun River in the mountains of Persia and traveled by boat up the Yangtze River in China.

Rough and Royal

When this British explorer was on the move, she both roughed it and traveled in luxury. She hung out with ordinary people as well as kings and queens. She wrote dozens of books about her adventures, and in 1890 she became the first woman **fellow** of the Royal Geographical Society. That was quite an achievement. At the time, the society had a reputation for ignoring women's exploration.

Traveling by camel caravan through Morocco, Isabella met and lived with the **Berber** people.

In her book The Yangtze Valley and Beyond, Isabella shared her adventures traveling up China's Yangtze River from the city of Shanghai.

Junko Tabai, Mountain Conqueror

Junko Tabai was born in Miharu, Japan, on September 22, 1939. She was a small and sickly child, but with seven children to support, her parents did not have the money to pay for sports and hobbies that could improve their daughter's health. Despite all of this, Junko went on to become one of the greatest mountain explorers in the world. And it all happened because of a class trip when she was 10!

Catching the Climbing Bug

Junko's first climbing adventure was to Mount Nasu, a chain of volcanoes in a national park. After that, she was hooked. But climbing was an expensive hobby, so Junko had to wait until she was at university to really get into it. There, she found out she couldn't join a local climbing club because it was for men only. So Junko formed her own club. She then went on to climb some of the world's toughest mountains.

Junko became a very skilled climber. Many climbers who were not as capable as Junko were able to get sponsorships to pay for their climbing because they were men. Junko had to work twice as hard as them, with equipment that wasn't as good, to get a sponsorship deal.

Junko had to make her own sleeping bags, gloves, and waterproof pouches because her sponsors didn't give her enough money to buy equipment.

Female Firsts

In 1975, Junko became the first woman to summit, or get to the top of, Mount Everest as co-leader of a women's-only climbing expedition. Junko and several of her team members nearly died on Everest when an **avalanche** hit their camp. Buried under snow for six minutes, she lost consciousness. Her **Sherpa** guides dug her out and after a few days' rest, she carried on climbing. Junko followed that by becoming the first woman to climb the Seven Summits, or the highest peaks on every continent.

Avalanches are a threat to climbers. Why do you think Junko carried on climbing despite the danger?

CLEANING THAT COUNTS

Junko didn't just climb mountains, she also helped cleaned them. She studied and wrote about how mountains were covered with garbage left behind by climbers. Junko also ran an organization that lead mountain cleanups—the Himalayan Adventure Trust of Japan.

This photo of Junko was taken in 1985. It shows her climbing Somoni Peak. It is the highest mountain in Tajikistan, Central Asia.

CHAPTER 4

AMAZING IN THE AMERICAS

Many women explorers were adventurers who traveled because they wanted to see new places. Others traveled because they were hired to do so—as guides, gatherers, or companions. These unsung heroes not only fed the men, but they also used their understanding of the land and the locals to help explorers reach their goals.

Sacagawea, Keeper of the Corps

From 1804 to 1806, a group of men were sent by President Thomas Jefferson to explore and map the western half of the United States. The project was known as "The Corps of Discovery Expedition," but was also known as the "Lewis and Clark Expedition." It was so-named for its leaders, Captain Meriwether Lewis and Second Lieutenant William Clark.

Far More than Just a Wife

The expedition was to take the group into parts of the United States never before traveled to by white men. To make this possible, Lewis and Clark hired a French-Canadian fur trader named Toussaint Charbonneau to act as a guide and **interpreter**. His wife, an Indigenous woman named Sacagawea, came along with him. Sacagawea would prove extremely useful to the group. In fact, Lewis and Clark soon learned that she was far more important than her husband.

Sacagawea's baby, Jean Baptiste, was the youngest member of the expedition. William Clark nicknamed him "Little Pomp." Sacagawea traveled through the entire expedition while carrying her baby boy strapped to her back. Why do you think this may have helped earn her the respect of the Corps?

Silver-Tongued Sacagawea

Sacagawea spoke two Indigenous languages, Shoshone and Hidatsa. This meant that she could communicate with both tribes whenever the party met them. Also, her presence in the group of white men helped reassure some Indigenous peoples that the Corps wasn't a threat, because for many Indigenous cultures, women rarely traveled with men who were going to war.

This painting shows Lewis and Clark with Sacagawea during their journey.

HARD WORK, NO PAY

As a child, 12-year-old Sacagawea had been kidnapped by Hidatsa warriors from her homelands near the Rocky Mountains. She was taken to the Hidatsa-Mandan villages (near Bismarck, North Dakota), and eventually married off to Toussaint. But she never forgot her home. Her memory of trails, or paths, she took during her childhood near the Rockies helped the Corps of Discovery through the difficult Bozeman Pass in Montana. Sacagawea was so important to Lewis and Clark that Clark called her their "pilot," yet when the expedition ended, Toussaint was paid for the job and given land. Sacagawea got nothing.

Sacagawea's knowledge of the western American landscape, particularly its mountains, helped the Corps survive.

Best Woman for the Job

Throughout the journey, Sacagawea proved that she was smart and quick-thinking. When one of the boats overturned, she quickly rescued its contents from the water. She also searched the plains for wild artichokes, plants that kept the Corps men alive at a time when they had run out of food and were starving. Sacagawea did all of this while caring for a newborn baby!

Together Again

In an incredible story of luck, Sacagawea met her long-lost brother on the journey west. The Corps had been traveling through Shoshone lands, hoping to trade for horses, when the group came across a small party that included Sacagawea's brother Cameahwait. The pair was full of joy to see each other again and with the help of her brother, Sacagawea traded for horses that would help the men cross the difficult Rocky Mountains.

Sacagawea collected herbs and other plants that added to the Corps's diet of meat. They also ate camas roots (right).

Today, Sacagawea is recognized as the true heroine that she was. This statue of her shows her carrying her baby.

Sacagawea also asked a guide to help the Corps travel through the Bitterroot Range of the Rockies. It was a difficult journey through snow that took two weeks.

I Came Here for the Ocean

Sacagawea helped the Corps reach the Columbia River near the Pacific Ocean. There, the men settled for the winter in a building they called Fort Clatsop. But Sacagawea hadn't yet seen the ocean. When a group of Indigenous people told the Corps that a whale was stranded on a beach nearby, Sacagawea asked that she be taken to see it. She had, after all, traveled too long not to see the waters her fellow explorers were so desperate to reach.

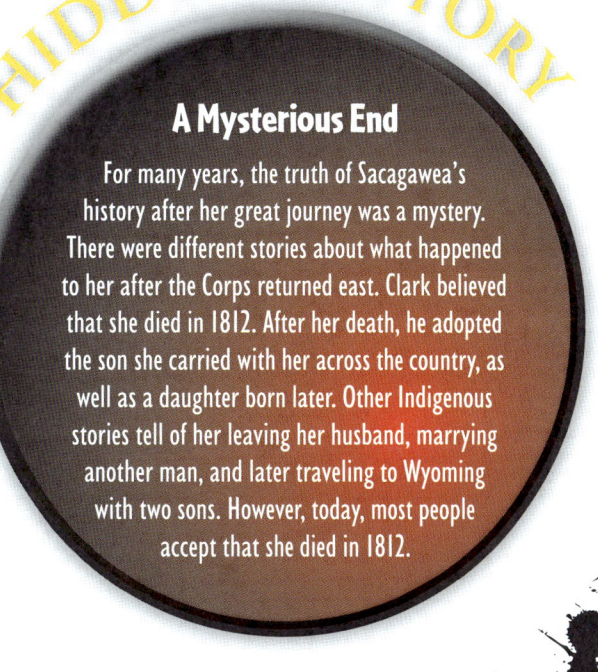

HIDDEN HISTORY

A Mysterious End

For many years, the truth of Sacagawea's history after her great journey was a mystery. There were different stories about what happened to her after the Corps returned east. Clark believed that she died in 1812. After her death, he adopted the son she carried with her across the country, as well as a daughter born later. Other Indigenous stories tell of her leaving her husband, marrying another man, and later traveling to Wyoming with two sons. However, today, most people accept that she died in 1812.

Octavie Renard Coudreau, Amazon Adventurer

The Amazon rain forest wasn't exactly a paradise for nineteenth-century Europeans used to comfortable living. It was sticky and hot, had swarms of disease-carrying mosquitoes, deadly snakes, and killer spiders. French explorer Henri Coudreau called it a "green Hell." But for Coudreau's wife and fellow explorer, Octavie Renard Coudreau, the thick rain forest and wild rivers became her life's work.

Not Beaten by the Rain Forest

Octavie spent seven years in the rain forest and on the Amazon River, yet said the place wasn't for her. Her work started with Henri, a geographer who spent years living with Indigenous Amazonian tribes.

Octavie's husband, Henri Coudrea (middle), was fascinated by the Amazon and its mysterious tribes.

Octavie wrote about the land, plants, and animals of the Amazon. Her detailed information was used by later explorers.

Not long after they married in 1899, the pair explored and mapped the Trombetas River, which fed into the Amazon River. However, Henri became sick with malaria and died in Octavie's arms. An ocean away from friends and family, Octavie made a coffin from boat planks. Then she buried her husband near a lake and went right back to exploring.

Never for Glory

Octavie's goal wasn't fame or glory. She also claimed it wasn't "for the love of geography." She said she wanted to finish the job that her husband had started so that she could dig up his body and return it home. And she did—enduring many hardships along the way. But as a grieving widow in the middle of the rain forest, she found herself with more power to follow her own path than she ever had in France.

THE PULL OF THE RAIN FOREST

During her years in the rain forest, Octavie was made an **official** explorer for the government of France. It was a position never before given to a woman. Octavie seemed to have mixed feelings about exploration. She wrote several books on her travels and also finished writing Henri's book *Voyage au Trombetas*. But after her role as official explorer ended, she never again went back to the rain forest.

The thick rain forest of the Amazon had many dangers, including diseases such as malaria and attacks by jaguars.

Charlotte Small, Country-Born Explorer

The fur trade led to a lot of exploration in North America. For more than 250 years, from the early 1600s to the mid-1800s, explorers working for fur-trading companies pushed farther into the continent in the relentless search for more furs and more fur-trading lands. Most of these explorer-traders were men. Charlotte Small was an exception to this rule.

Small but Mighty

Tiny 5-foot (1.5 m) tall Charlotte was the **Métis** daughter of a Cree woman and a white fur trader. She herself then married a white fur trader and explorer, David Thompson, and traveled with him on several expeditions. Charlotte's language skills and ability to trap animals, cook wild foods, and make clothing out of animal skins, made her hugely important on their long explorations through western Canada, the Rockies, and into the northwest United States.

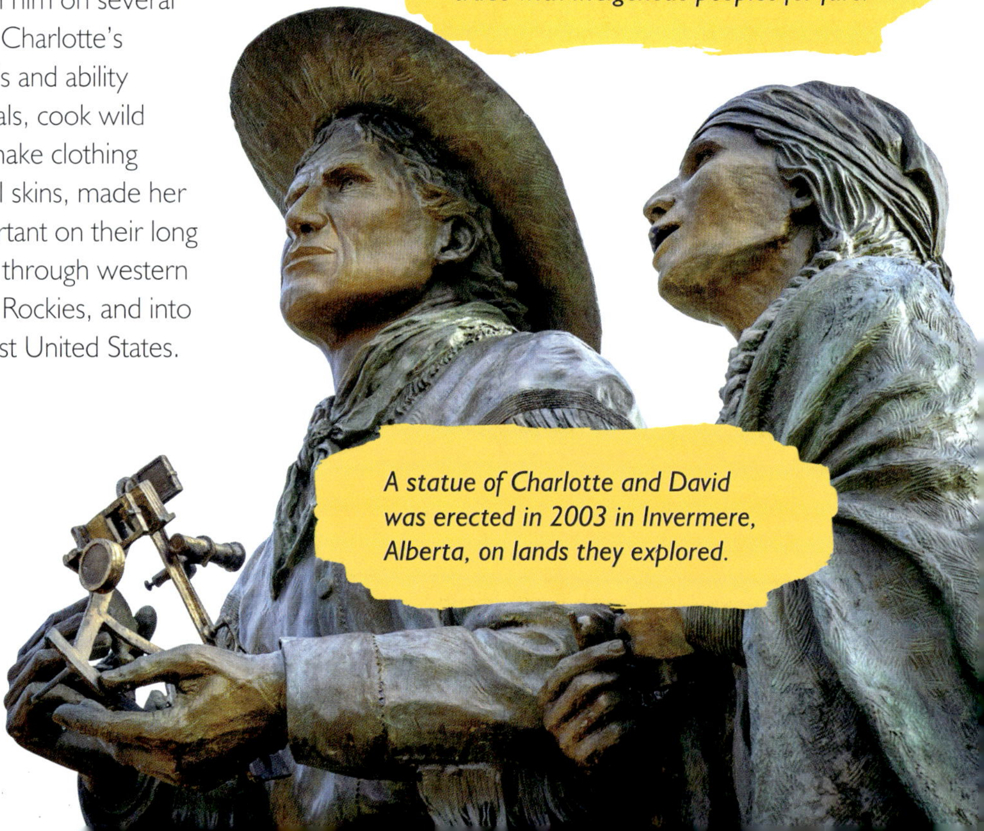

The Hudson Bay Company (HBC) was set up in 1670 and had trading posts throughout what is now western Canada. Trading posts were places where company men worked and goods were brought to trade with Indigenous peoples for furs.

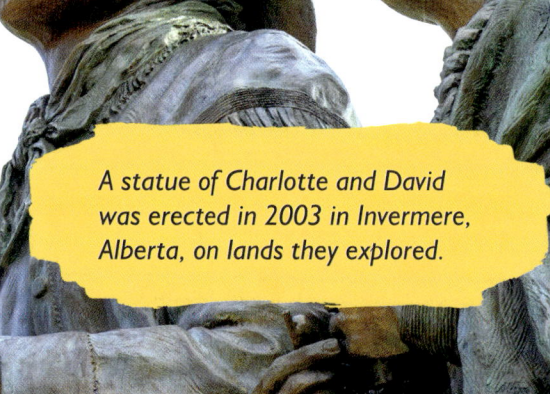

A statue of Charlotte and David was erected in 2003 in Invermere, Alberta, on lands they explored.

CHILD BRIDE

Charlotte was just 14 when she agreed to be David's wife. Her Cree family set up the match, because Charlotte's European father, Patrick, had earlier abandoned his family and returned to Europe. Charlotte's mother was a "country wife." These were Indigenous women who had been married to European traders in Indigenous ceremonies. Their children became the Métis people. Unlike many country marriages, Charlotte's lasted 58 years, until David's death.

Charlotte the Super Mom

Charlotte did everything the men in David's exploration crews did, and more. She traveled 12,000 miles (19,312 km) on foot, horseback, by snowshoe, dog sled, and canoe. She also did this with three of the couple's children in tow—giving birth to two more during treks through the wilderness.

Traveling through the Rockies was dangerous and took a long time. On one trek, a scared horse nearly killed two of Charlotte's children. On another journey, her young daughter got lost when she was separated from the group. Luckily, the group managed to find her.

Settling Down

When David retired from the fur trade, he took Charlotte and the children and settled down in Quebec. They lived there for the rest of their lives, with Charlotte dying just a few months after her husband.

David knew marrying Charlotte would help his career and give him an advantage over other traders. She could speak and translate Cree, English, and French with the Indigenous trappers and hunters she and David met on their travels.

Ada Blackjack, Born Survivor

The only survivor of a 1921 **Arctic** exploration team, Ada Blackjack spent two years on an extremely cold and distant Arctic island before being rescued in August 1923. She lived thanks to her determination and incredible survival skills.

There to Work

Ada was not a **traditional** explorer. She did not seek gold, glory, or a place in history. She was in the Arctic to do a job for which she had the skills. An **Inupiat** woman from Alaska, Ada was hired to cook and sew for the men on an expedition. The expedition was carried out to claim Wrangel Island in the Chukchi Sea for Canada.

Ada had hunted, fished, and lived off the land since she was a child. Her ability to find and kill wild animals, using their meat for food and skins for clothing, helped her survive when the men of the expedition all died.

In Alaska, Caribou and reindeer are important herd animals for their meat and skins.

A relief ship was supposed to drop off supplies to the expedition on Wrangel Island (shown right) after a year, but it never arrived.

Only Ada

The expedition was made up of four men and Ada. The group took enough supplies to last them only six months on the island, believing that a relief ship would arrive to bring them more at the end of the summer. When the ship did not come, the men realized they would have to survive on the island by hunting for food. However, with little to eat, they grew weak and had difficulty hunting. While Ada cared for one dying man, three others headed south to find help—and were never seen again. After the man she was caring for died, Ada survived several months on her own before a rescue ship arrived and took her home.

SINGLE MOM SURVIVOR

Ada joined the expedition to the Arctic after her husband died, leaving her a widow with a son. She survived on Wrangel Island by trapping animals and using a rifle left behind by the men to shoot seals. Using Indigenous knowledge, she chewed the animal skins and made them into boots. After her rescue, Ada moved back to Alaska and worked herding reindeer. She died at 85, having passed on her hunting and trapping skills to her children.

Bessie Coleman, Sky-High Heroine

The daughter of a Texan farmer, Bessie Coleman's adventurous spirit led her to become the first woman of both African American and Indigenous descent to hold a pilot's **license**. Bessie faced prejudice and hatred along the way, but she never let that hold her back...

Fly-Girl Fantasies

Living in Chicago in 1916, Bessie had heard stories of World War I flying aces and their amazing skills. She decided she wanted to fly too. Bessie saved for flight school, but not one school in her home country would agree to train an African American woman as a pilot. Hearing of her goal, an African American newspaper publisher suggested she go to France to get her license, where it was easier for African Americans to train to fly. Bessie then learned French, went to France, and earned her pilot's license.

Bessie the Barnstormer

Bessie took her pilot training back to North America. Before long, she became a **barnstorming** stunt pilot and appeared in shows throughout the country.

After her training in France, Bessie found she couldn't get enough work as a pilot in the United States. She then took further training with famous flying aces in France, Germany, the Netherlands, and Switzerland.

Bessie's stunts were all anyone could talk about wherever she went. She gave lectures about her adventures. She also gave flying lessons to African Americans all across the country. Bessie's dream was to eventually own her own plane and open a flying school for African Americans.

A Dangerous Life

The life of a high-flying adventurer was exciting, but also very dangerous. Bessie survived a major crash in 1923 that left her with a broken leg. Still, she kept flying. Her flying days ended in 1931, when Bessie fell out of the passenger seat during a test flight in which her mechanic flew the plane. Sadly, Bessie died. Every year on the date of her death, African American pilots fly over and drop flowers onto her grave in Chicago's Lincoln Cemetery in memory of this amazing adventurer.

STANDING FIRM

Bessie used her appearance fees from stunting across the country to buy her own plane. It gave her the freedom to say when and where she would perform in front of large crowds. At one appearance in her Texas hometown, Bessie was told the entrance for her audiences would be segregated, or separated, so white people would not have to enter with black people. Bessie refused to perform unless there was one gate for all. Bessie got her way.

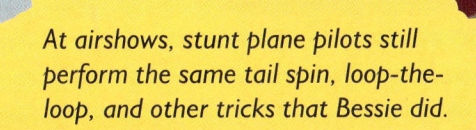

At airshows, stunt plane pilots still perform the same tail spin, loop-the-loop, and other tricks that Bessie did.

CHAPTER 5

ACING IT IN AFRICA

Some women explorers were striking in their fearlessness. Not only did they pluck up the courage to travel to strange lands, but many also stood up to wrongdoing they witnessed along the way. This included fighting against slavery and other injustices and cruel treatment.

Florence von Sass Baker, from Slave to Explorer

Despite starting out life as a slave, Florence von Sass Baker went on to become an explorer, slave-trade fighter, and a lady. Born in Transylvania (now Romania) in 1841, Florence was kidnapped as a child and grew up in a **harem**. At 14, she was sold as a slave in the city of Vidin, Bulgaria. She might have ended her days in slavery, except for chance. English explorer Samuel Baker happened to be in Vidin and saw her being sold at an auction. Determined to free her, he **bid** on her. When that failed, he paid a bribe to free her.

Florence became a fierce freedom fighter after Samuel freed her and once even threatened slave traders with her umbrella!

Florence spoke several languages, including English, Turkish, and Arabic.

40

Florence and Samuel saw many amazing animals on their journey along the Nile River, including hippos and crocodiles.

Sailing on the Nile

In 1861, Samuel took Florence to East Africa where the pair searched for the source of the Nile. They sailed up the river, before traveling on foot, covering more than 500 miles (805 km). Although Florence didn't find the source of the Nile on the journey, she did find an important job. During her travels she had passed through the slave-trading center of Gondokoro, in South Sudan. Florence was horrified by the slavery she saw there and vowed to find a way to fight it.

Florence the Fighter

Next, Florence and Samuel went to England and married. They returned to Gondokoro in 1869, where Samuel was given the job of leading a military expedition to fight the slave trade. In one battle in Bunyoro, Uganda, Florence fought slave traders with rifles, a pistol, and her umbrella. After four years of fighting, Samuel's appointment came to an end. He and Florence then headed back to England.

HIDDEN HISTORY

Keeping Secrets

Florence and Samuel hid the story of how they met for many years. They knew that people might not take it well if they learned Samuel tried to buy Florence as a slave, then bought her freedom and eventually married her. When Samuel was **knighted** in 1865, Florence became Lady Baker, but she was never quite accepted by society. People did not like the fact that Florence had been a slave and then made a lady.

Isabelle Eberhardt, the Wandering Writer

Isabelle Eberhardt was an extraordinary woman for her time. Born in 1877, she dressed like a man, spoke seven languages, and generally did what she wanted! Isabelle said her travels across the desert in Africa were just "obeying her destiny." Her destiny, or future, also included writing travelogues, or journals of her travels, moving to Algeria, learning Arabic, and marrying an Algerian soldier.

All-in Isabelle

Isabelle did not do things halfway. When she signed up to an adventure or cause, she was all-in. Many of the things known about her through her journals are questioned, because she constantly made up stories. Isabelle was a skilled writer and bold thinker, and she sometimes used her talents in her explorations. When she was just 20, she moved to Algeria, Africa, which was under French rule. There, Isabelle wore her hair short, dressed as a man, and wandered from place to place. She had French-Algerian society gossiping by marrying an Algerian soldier and then staying apart from him for periods of time.

In Isabelle's travelogues, published after her death, she talks about how she fell in love with North Africa.

Sometimes Isabelle traveled alone and at other times with a guide, who would have also cooked on the journey.

42

Isabelle dressed as a man in Europe and wore traditional Algerian clothing in the deserts of North Africa.

Super Storyteller

To make a living, Isabelle wrote for a newspaper that sent her to write about things throughout the country. For a time, she lived in the Sahara among the **Bedouin** people. Her stories about their life and culture were published in the paper. Isabelle's writing became known after her death when her publisher friend took her travelogues and published them as books. In one book, *The Oblivion Seekers*, Isabelle said of herself "No one ever lived more from day to day or was more dependent upon chance."

Writing from the Battlefield

Isabelle also reported on a battle fought between the French and Moroccan tribesmen. She lived among the soldiers, like modern journalists do. Some say she was even the first war **correspondent**. Isabelle developed a sympathy for the Algerian people, changed her name to Si Mahmoud Saadi, and converted to Islam, the religion of Muslims.

SWEPT AWAY

After her death in a flood at the age of 27, Isabelle's travelogues were published. In 1991, they were made into a movie about the life of this extraordinary woman.

Mary Kingsley, Adventurer of Africa

Mary Kingsley didn't seem like an adventurer when she was young. She was a caring daughter who tended to her seriously sick parents. But when they died, leaving her an **inheritance**, Mary let loose and did what she always wanted to do: travel.

Mary on a Mission

From her home in England, Mary set off on her first journey to Africa in 1893. She was 30 years old, unmarried, and traveling alone. Most of the other women sailing to Africa at that time were Christian **missionaries** traveling with their husbands. Mary had a different mission. She was hoping to finish a book on African cultures that her father, a doctor and explorer, had started years earlier.

Despite being so adventurous, Mary carried out all her activities wearing traditional clothing!

CORSETED CLIMBER

For all her open-mindedness, Mary's clothes were very traditional. Even in the exhausting heat, she wore a buttoned up white shirt and long dark skirt. Despite her hot and heavy clothing, when her companions collapsed from the effort of climbing Mount Cameroon, Mary made it up the 13,760-foot (4,194 m) mountain in a chest-crushing corset!

Mary's first trip wouldn't be her last. All in all, she journeyed to Africa three times, visiting Angola, Sierra Leone, Niger, Nigeria, Equatorial Guinea, French Congo, Gabon, Cameroon, and South Africa.

A Strange Sight

A white British woman trekking through jungles, fishing in swamps, and collecting insects was an odd sight at the time. Mary was often asked why her husband wasn't with her on her travels. But exploring alone had its advantages. She hired her own guides and lived among local people, from whom she learned much.

Mary worked as a nurse during the second Boer War (1899 to 1902) in South Africa. In 1900, she died from an illness caught while nursing soldiers.

Freethinking Mary

In Mary's day, most European explorers of Africa saw the continent as backward, with cultures inferior to their own. They had narrow ways of thinking and superior attitudes. Mary was different. She was interested in the daily life of the Africans she met. She even lived with the Fang people of Equatorial Guinea, whom other European explorers thought were **cannibals** because they kept human bones. But, as Mary discovered, the bones were in fact those of their **ancestors**. The Fang people kept the bones as part of their religion and so they could stay close to their dead ancestors.

Mary is remembered here on this 2008 stamp from the Democratic Republic of Congo (DRC).

GLOSSARY

abandoned Left behind
adopted Legally made the child of of someone other than a biological, or natural, parent
Alps A mountain range in southern Europe
ancestors The people who a person is descended from, such as great grandparents
archaeologists People who study ancient cultures by examining buildings and objects that people who lived long ago have left behind
Arctic The most northerly region of the world
assassins People who kill others, often for political reasons
astronauts People trained for spaceflight
authorities People who have power to make decisions and enforce rules and laws
avalanche A large amount of snow that suddenly falls down a mountain slope
barnstorming To stunt-pilot a plane in farming and rural areas
Bedouin An Arabic people who travel through and live in the desert
Berber Desert people who travel throughout North Africa
bid To place a bet, or an amount of money, in the hope of winning something
bloomers Frilly trouserlike clothing worn by women under their skirts
cannibals People who eat human flesh
capsules Parts of spacecraft in which astronauts land back on Earth
claim To make a demand or statement, or to take something and say it is yours
convent A community of nuns and the buildings in which they live
correspondent A person who reports on events from a foreign country
culture A large group of people who share the same language and ideas
Dalai Lamas The spiritual leaders in Tibetan Buddhism
debts Money that is owed to other people
earl A British nobleman, or member of the upper class
fellow A member of a group or organization

gender How a person identifies, such as male or female
harem A group of women who all have a relationship with one man, usually a king
Hindu Related to Hinduism, a religion of India
Indigenous The original inhabitants of a region or country
inheritance Something that is passed down to the relatives of a dead person, such as money
interpreter A person who translates the words that someone is speaking into a different language
Inupiat A group of Arctic peoples and their language
Kazakhstan border The border of Kazakhstan, a country in central Asia
knighted Someone who is made a knight by a king or queen
Lady The female equivalent of a lord
license Permission or a permit to do something
medieval Referring to the time in history between the fifth and fifteenth centuries
Métis A person of mixed Indigenous and European ancestry
Minaret of Jam A mosque tower in western Afghanistan that was built around 1190 C.E.
missionaries People who went to non-Christian countries to spread Christianity
monastery A community of monks and the buildings in which they live
naturalist A person who studies plants and animals
official A person who is given the job of performing special duties
orbit To move around Earth
orphan A person whose parents are dead
orphanage A place were orphans are cared for
palm wreath A crown made out of leaves
pension Money paid to a person once they retire, or stop working
pilgrimage A journey to a religious place
prejudice An opinion about someone or something that is formed without knowledge
race A group of people who are all related to a

distant ancestor
records Events that are set down in writing
satellite A humanmade object that orbits Earth and is used to send information to Earth
service module The part of a spacecraft that contains the systems it needs to work in space
Sherpa A member of a Himalayan people who are known for their skill in mountaineering
slavery The practice of buying and selling people as property, and forcing them to provide free labor
Soviet Union Before 1991, the name given to the country that is today called Russia
spindles Round rods on a spinning wheel

sponsorship deal Funding for a person or athlete
territory An area of land belonging to a country
Tibetan Buddhism The type of Buddhism practiced in Tibet
traders People who buy and sell goods
traditional Describes ways of doing things that have been carried out for hundreds of years
translated Changed from one language to another
tribesmen People who belong to the same tribe, or group
turban A head covering that is made of a long cloth wrapped around the head
zero gravity Weightlessness in space

LEARNING MORE

Read more about women who broke the rules and changed the world!

Books

Cooke, Tim. *Explore with Mary Kingsley* (Travel with the Great Explorers). Crabtree Publishing, 2018.

Cummins, Julia. *Women Explorers: Perils, Pistols, and Petticoats!* Puffin Books, 2015.

Dakers, Diane. *Amelia Earhart: Pioneering Aviator and Force for Women's Rights* (Crabtree Groundbreaker Biographies). Crabtree Publishing, 2016.

Websites

Read Bessie Coleman's story at:
www.nationalaviation.org/our-enshrinees/coleman-bessie

Could an all-female space mission be the next amazing adventure? Find out at:
www.nationalgeographic.com/magazine/2019/07/space-travel-four-ways-women-are-a-better-fit-than-men

Read the biographies of other women explorers at:
www.nationalgeographic.com/news/2018/04/female-explorers-adventure-marriage-career

INDEX

Africa 4, 6, 7, 21, 24, 40–45
Amazon rain forest 32
Amazon River 32
Americas 28–39
Asia 7, 8, 18–27

Baré (Baret), Jean 8–9
Bird, Isabella 24–25
Blackjack, Ada 36–37
Boer War 45

Canada 5, 34, 36
Charbonneau, Toussaint 28, 29
Cohen Kopchovsky, Annie 16
Coleman, Bessie 38–39
Commerson, Philibert 8–9
Corps of Discovery Expedition 28–31

David-Néel, Alexandra 20–21
desert 4, 10–13, 18, 42–43
Dido 6–7
Dutch East India Company 4

Eberhardt, Isabelle 42–43
Europe 7, 8–17, 22–24, 32, 35, 43, 45

fur trading 28, 34

guides 18–23, 27, 28, 31, 42, 45

Hudson Bay Company (HBC) 34

Indigenous peoples 7, 28–29, 31–32, 34–35, 37–38

Kingsley, Mary 44–45

Londonderry, Annie 16–17

Métis people 34–35
Middle East 10–11, 18, 21
missionaries 44
mountains 19–20, 22, 24–27, 29–30, 31, 44

North America 5, 24, 34, 38

Polo, Marco 18

Renard Coudreau, Octavie 32–33
rivers 25, 31–33, 41

Sacagawea 28–31
Small, Charlotte 34–35
Stanhope, Lady Hester 10–13
Stark, Freya 18–19
Swale Pope, Rosie 4

Tabai, Junko 26–27
Tereshkova, Valentina 14–15
Thompson, David 34

Vikings 4–5
Virgil 6–7
von Sass Baker, Florence 40–41

World War I 18, 23, 38
World War II 23

ABOUT THE AUTHOR

One of Ellen Rodger's first memories is of trying to explore her neighborhood block when she was three years old. She got lost and was brought home by the local police. Since then, she's walked on the Columbia Icefield in Alberta, Canada, traversed the Hussaini bridge in Gilgit-Baltistan, Pakistan (the rickety old one), and visited a farm school in rural South Africa.